I0795307

BRAVE THE BIOME

ISLAND SURVIVAL GUIDE

HEATHER C. HUDAK

CRABTREE
PUBLISHING COMPANY
WWW.CRABTREEBOOKS.COM

BRAVE THE BIOME

Author: Heather C. Hudak

Editors: Sarah Eason, Jennifer Sanderson, and Ellen Rodger

Proofreader and indexer: Tracey Kelly

Proofreader: Petrice Custance

Editorial director: Kathy Middleton

Design: Jessica Moon

Cover design: Tammy McGarr

Photo research: Rachel Blount

Production coordinator and Prepress technician: Tammy McGarr

Print coordinator: Katherine Berti

Consultant: David Hawksett

Produced for Crabtree Publishing by Calcium

Photo Credits:
t=Top, c=Center, b=Bottom, l= Left, r=Right

Inside: Jessica Moon: p. 29t; Shutterstock: aSuruwataRi: p. 25t; BlueOrange Studio: pp. 12–13; Eric Buermeyer: pp. 20–21; Dmytro Buianskyi: p. 22b; Rich Carey: p. 28b; Cat Scape: p. 16b; Divedog: p. 17t; Everett Historical: pp. 8–9; Flystock: pp. 30-31; FOTOGRIN: p. 4; Golden Shrimp: p. 23b; Daria Iva: p. 13b; Kristopher Kettner: p. 5; Yevgen Kravchenko: p. 11b; Louella938: p. 21r; Joyce Mar: p. 11t; Maridav: p. 12b; Mbrand85: pp. 26–27; MikeDrago.cz: pp. 10–11; Dudarev Mikhail: p. 14b; Nikolaev Mikhail: p. 22t; Motortion Films: pp. 6–7, 24–25, 29b; Multiverse: pp. 28–29; New Africa: p. 7t; Kylie Nicholson: p. 32; Anastasia Nio: p. 17b; Tamara Obodieieva: pp. 22–23; Alena Ozerova: p. 13t; Vera Petruk: p. 25b; Ondrej Prosicky: p. 19t; Jason Patrick Ross: pp. 14–15; Alexey Seafarer: p. 26b; Satit Sewtiw: pp. 18–19bg; Jo Ann Snover: pp. 16–17; Bodor Tivadar: p. 6b; Vectorgoods Studio: p. 19b; Wildestanimal: pp. 18–19.

Cover: All images from Shutterstock

Library and Archives Canada Cataloguing in Publication

Title: Island survival guide / Heather C. Hudak.
Names: Hudak, Heather C., 1975- author.
Description: Series statement: Brave the biome | Includes index.
Identifiers: Canadiana (print) 20200285610 |
Canadiana (ebook) 20200285696 |
ISBN 9780778781332 (softcover) |
ISBN 9780778781271 (hardcover) |
ISBN 9781427125736 (HTML)
Subjects: LCSH: Shipwreck survival—Juvenile literature. |
LCSH: Survival—Juvenile literature. |
LCSH: Islands—Juvenile literature.
Classification: LCC GV200.5 .H83 2021 | DDC j613.6/909142—dc23

Library of Congress Cataloging-in-Publication Data

Names: Hudak, Heather C., 1975- author.
Title: Island survival guide / Heather C. Hudak.
Description: New York : Crabtree Publishing Company, [2021] |
Series: Brave the biome | Includes index.
Identifiers: LCCN 2020029948 (print) | LCCN 2020029949 (ebook) |
ISBN 9780778781271 (hardcover) |
ISBN 9780778781332 (paperback) |
ISBN 9781427125736 (ebook)
Subjects: LCSH: Shipwreck survival--Juvenile literature. |
Wilderness survival--Juvenile literature. |
Islands--Juvenile literature.
Classification: LCC G530.S42 H83 2021 (print) |
LCC G530.S42 (ebook) | DDC 613.6/909142--dc23
LC record available at https://lccn.loc.gov/2020029948
LC ebook record available at https://lccn.loc.gov/2020029949

Crabtree Publishing Company
www.crabtreebooks.com 1-800-387-7650

Printed in the U.S.A./092020/CG20200810

Published in Canada
Crabtree Publishing
616 Welland Ave.
St. Catharines, Ontario
L2M 5V6

Published in the United States
Crabtree Publishing
347 Fifth Ave.
Suite 1402-145
New York, NY 10016

Published in the United Kingdom
Crabtree Publishing
Maritime House
Basin Road North, Hove
BN41 1WR

Published in Australia
Crabtree Publishing
3 Charles Street
Coburg North
VIC, 3058

CONTENTS

Disclaimer

The information given in this book is advice on how to survive in the wilderness. No one should deliberately put themselves into any dangerous or extreme situations to carry out the activities. Everyone should follow expert advice before embarking on a journey in the wild. The advice given here is a guide only, and the publisher cannot accept responsibility for any injuries, loss, damage, or prosecutions resulting from the information given in this book.

IN THE MIDDLE OF THE OCEAN

If a boat capsizes or an airplane crashes in the ocean, survivors may find themselves alone on a deserted island. Deserted islands are sometimes called desert islands. These islands are found in remote places. There may not be anyone on the island to help survivors find their way home. Knowing basic survival skills can keep people stranded on an island safe until help arrives.

FAR FROM CIVILIZATION

Many islands have never been **inhabited** at all. Others have been deserted, such as St. Kilda in Scotland, or Dry Tortugas in Key West, Florida. This means that people lived there at one time, but they no longer do so. It is important for those stranded on an island to keep calm, focused, and positive. Islands often have a lot of food and materials for shelter and fire. They are surrounded by fish-filled waters, and there are very few large **predators**.

Ships and planes often carry **flares** that create a burst of light or cloud of smoke to signal for help.

HOT OR COLD

People often picture sunny, warm places when they think of islands, but this is not always the case. Islands are found in all parts of the world, from the cold north to hot **tropical** climates. Devon Island in northern Canada is the largest uninhabited island in the world. It is a cold, dry, rocky place where the ground remains frozen most of the year. Few plants and animals can survive in these harsh conditions. The Rock Islands of Palau in the Pacific Ocean are known for their tropical rain forests, sandy beaches, and coral reefs.

This boat has been shipwrecked on a tropical island.

LOOK FOR ...

Look for the "How to Survive" and "Be Prepared" features in this book. These list many of the techniques that people have used to survive on islands.

PREPARE TO SURVIVE

Most people do not prepare to be stranded on an island. No one expects their plane to crash or boat to sink. But if they have basic survival skills, they can make use of materials they find around them to keep safe until help arrives. The first thing to do when stranded on an island is to look for other survivors. It is easier for people to cope when they know they have support.

KEEP IT COOL

When stranded on an island, survivors need to do their best to keep calm. This lets them think clearly, so they can come up with a plan. When people panic, they are more likely to make poor choices. Being able to **adapt** quickly to any situation is an important survival skill. From building a fire to looking for food, survivors need to do whatever it takes to stay alive.

BE PREPARED

While searching the wreckage, survivors should wear sturdy shoes or boots that will protect their feet from poisonous sea creatures, such as sea urchins. Sandals and sneakers do not provide good protection from **hazards** but are better than nothing.

CHECK THE WRECKAGE

Survivors can check the wreckage for items they can **salvage** and reuse while waiting for help to arrive. Water, food, fire, and shelter are the four things people need to survive. There is a good chance of finding bottled water among the wreckage or at least containers that can be used to collect water. There may also be prepackaged foods and a first aid kit. Pieces of the boat or plane can be used to make a shelter. A battery and fuel can be used to start a fire. Blankets, plastic bags, rope, a knife, flashlights, a lighter, and clothing can also be useful.

The life jackets or seat cushions survivors use for floating as they swim to shore can also be used for bedding on land.

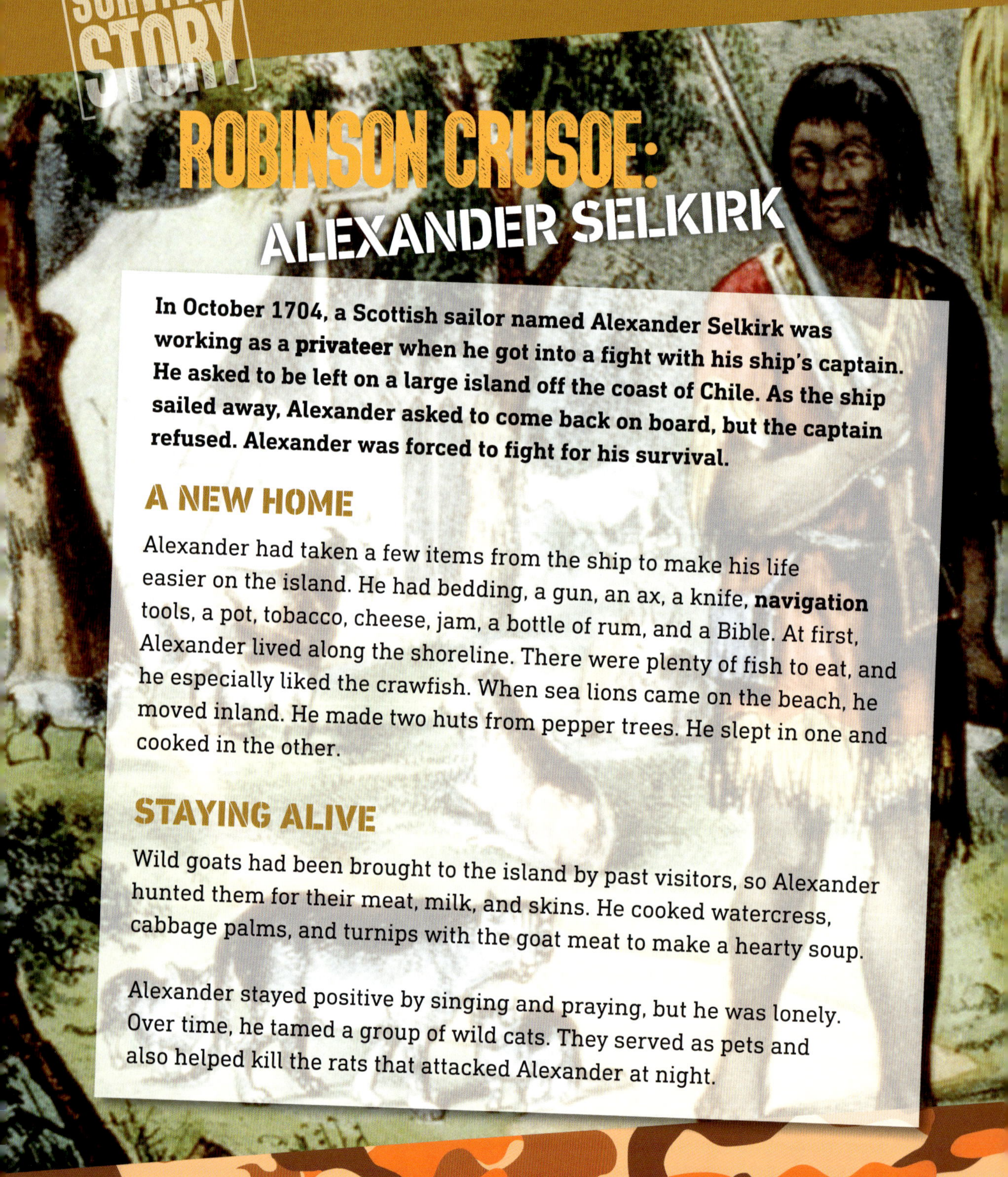

SURVIVAL STORY

ROBINSON CRUSOE: ALEXANDER SELKIRK

In October 1704, a Scottish sailor named Alexander Selkirk was working as a privateer when he got into a fight with his ship's captain. He asked to be left on a large island off the coast of Chile. As the ship sailed away, Alexander asked to come back on board, but the captain refused. Alexander was forced to fight for his survival.

A NEW HOME

Alexander had taken a few items from the ship to make his life easier on the island. He had bedding, a gun, an ax, a knife, **navigation** tools, a pot, tobacco, cheese, jam, a bottle of rum, and a Bible. At first, Alexander lived along the shoreline. There were plenty of fish to eat, and he especially liked the crawfish. When sea lions came on the beach, he moved inland. He made two huts from pepper trees. He slept in one and cooked in the other.

STAYING ALIVE

Wild goats had been brought to the island by past visitors, so Alexander hunted them for their meat, milk, and skins. He cooked watercress, cabbage palms, and turnips with the goat meat to make a hearty soup.

Alexander stayed positive by singing and praying, but he was lonely. Over time, he tamed a group of wild cats. They served as pets and also helped kill the rats that attacked Alexander at night.

YEARS PASSED

Alexander thought only a few days would pass before help arrived, but it was four and a half years before he was rescued. Finally, on February 2, 1709, a British ship arrived. Alexander told the crew his story of survival. One of the men had been on the boat that left Alexander behind. The man told Alexander he had been right to leave the ship—it sank not long after, and only a few sailors survived. These unfortunate men were then taken prisoner by the Spanish. Alexander left the island and joined the crew on their adventures over the next two years before returning home to Britain.

Many people believe English author Daniel Dafoe's famous 1719 book Robinson Crusoe *was based on Selkirk's real-life experiences on his island. This illustration shows Crusoe with a dog, an animal not featured in the book.*

BUILD A SHELTER

Shelter can protect people from insect bites and harsh weather, such as thunderstorms, wind, and extreme sunlight. Even on a tropical island, temperatures can drop a lot at night. It is important to keep warm to avoid getting hypothermia. A raised bed inside the shelter provides a safe place to rest. It helps keep insects, snakes, rodents, and other animals from crawling on people while they sleep.

IDEAL LOCATION

The best place to build a shelter is next to the shore where there are no big trees or plants to hide survivors from the view of a passing ship or plane. In hot places, being close to the water also makes it easy to go for a swim to cool off. It is best to build shelters away from trees with hanging fruit, such as coconuts, that could fall on people.

It can get cold when the Sun goes down, even in warm places. Building a small fire is one way to keep warm.

KEEP IT SIMPLE

Shelters such as a **lean-to** or a **tepee** are quick and easy to build. For a lean-to, a long, straight branch can be leaned against a tree at an angle. Walls can then be made by placing smaller branches at an angle along the larger branch. A tepee is made by sticking three tall branches into the ground in the shape of a **tripod**. Then, more tall branches can be leaned upright against the tripod to form a circle. The lean-to or tepee is covered with twigs, sticks, or leaves to keep out animals and bad weather.

lean-to shelter

HOW TO MAKE ROPE

Rope is very useful on an island. One of the easiest ways to make rope is from dead plant stalks. The first step is to break any leaves, flowers, or branches off the top of the stalk. Then, the stalk needs to be flattened to soften the wood inside. The next step is to snap off one end of the stalk. The **fiber** on the outside can then be peeled apart from the wood inside. Several pieces of fiber can be twisted or braided together to form a rope.

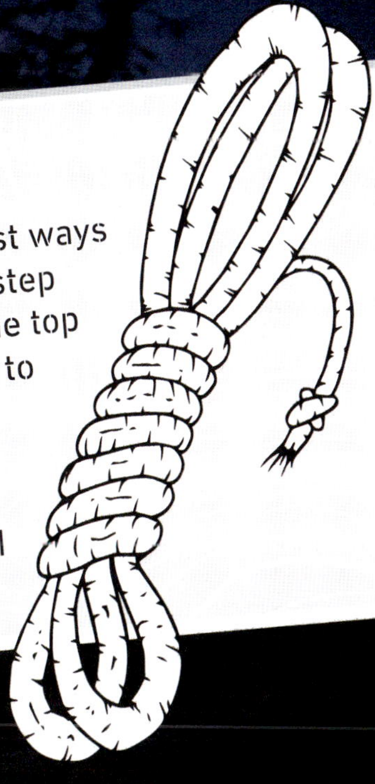

FIND WATER

Nothing is more important than finding a source of drinking water when stranded on an island. People can survive without food for up to three weeks, but they can last for only three or four days without water.

FRESHWATER SUPPLY

None of the water around an island is drinkable because the human body cannot process the huge amount of salt found in seawater. Drinking it causes **dehydration** and can lead to death. The beach is a good place to start when looking for **fresh water** to drink. There are often small streams or tiny flows that people can follow to the source of the water. Walking inland may lead to streams or springs. Water should be boiled before drinking. This removes any **bacteria** and **parasites** that can make people sick.

People can use large leaves to collect rainwater.

OTHER OPTIONS

Many plants have water in their leaves, stems, roots, and stalks. Vines, coconuts, cacti, and other plants or fruit can be cut open to find water inside. Water often drips from the ceiling of caves. Survivors can place a container underneath to catch it. Dew and rain are also great sources of clean water. A plastic bag can be hung from a tree to catch rainwater as it falls. Another option is to dig a hole in the ground and wait for it to slowly fill with water. A cloth can be used to soak up the water and wring it out into a container.

BE PREPARED

Objects wash up on islands all the time. These may include plastic bottles, cans, shoes, and other items that can be used to collect water.

SURVIVAL STORY

DYING OF THIRST: JOE AND LORENZO

In 2000, Joe Rangel and Lorenzo Madrid left for a fun-filled sportfishing trip in the Sea of Cortez off the coast of Mexico. They could not have imagined what would come next. They were stranded on a desolate, craggy island after their fishing boat became separated from their charter ship.

ADVENTURE AWAITS

On September 30, Joe and Lorenzo boarded a chartered fishing boat in San Felipe, Mexico, to begin their adventure. On their fourth day at sea, the men set out in a smaller boat to explore the area with a local guide. As night fell, it became clear that the guide had lost his way. The larger ship was nowhere in sight, and the sea started to get rough. They had no radio or flares to signal for help. There was no safety gear or life jackets on the boat. All they had were three bottles of beer, four bottles of water, a can of soda, and some dried fruit.

After only a few hours without water, people start to feel the effects of dehydration. They may not be able to think clearly or function properly.

ROCKY SHORE

The men headed for the nearest island and soon found themselves on Isla Ángel de la Guarda, about 200 miles (322 km) south of the U.S. border with Mexico. There is not much on the island other than rocks and cacti. After three days of being stranded on the island, help had still not arrived. The men decided to try to escape. They used pieces of wood to paddle to their boat, but they soon grew tired and drifted back to the island. Their boat smashed apart on the rocks as they did.

ONE MORE TRY

The men had little trouble finding food. They ate crabs, sea snails, and grasshoppers to keep their bellies full, but water was their biggest challenge. They drank from plastic bottles that had washed up on shore. They were so thirsty, but they only found about a third of a bottle of water each day. It was not enough. At night, they took shelter in caves to protect themselves from the strong winds on the island.

The harsh conditions were too much for Lorenzo, and he died from dehydration and **exposure**. Joe knew that it was only a matter of time before he and the guide met the same end. They built a wooden raft and drifted along the shoreline, so they would be easier to see. Finally, after being lost for 13 days, the two survivors were rescued.

HUNT FOR FOOD

People need food to keep up their energy. Some islands are covered in lush plants. Others are barren, or without plants that can be eaten, and have few signs of life. Still, it is usually possible to find some food to eat if people know what to look for.

START BY THE SEA

Crabs, sea urchins, mollusks, oysters, clams, and mussels can often be found in tidal pools and shallow waters along a beach. Some seafoods contain parasites or bacteria that can make people sick. If they are difficult to open, it is a sign that they are probably okay to eat. It is best to avoid eating any sea creatures that are covered in barbs, or spines, such as stonefish. They may be poisonous. All seafood should be cooked before eating it. Fresh seaweed also grows along the beach. It contains many vitamins and minerals but can also make people thirsty.

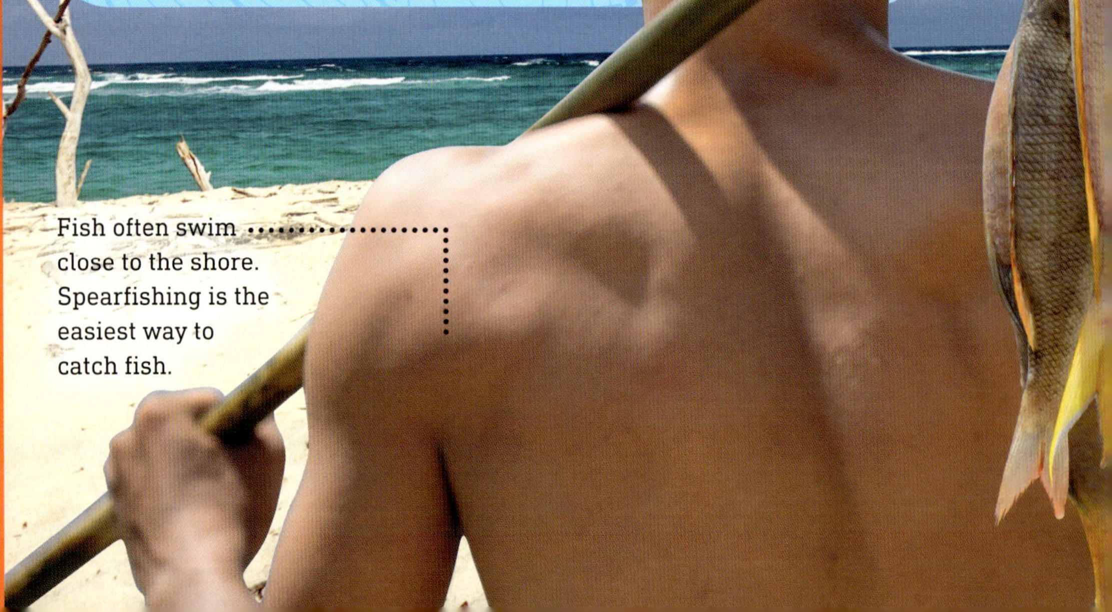

Fish often swim close to the shore. Spearfishing is the easiest way to catch fish.

SAFE TO EAT

People should never eat a plant if they are not sure what it is. It is safer to focus on looking for foods that are easy to recognize, such as coconuts, plantains, pineapples, and bananas. Many berries and mushrooms are poisonous. Plants that are rotten, have leaves that grow in groups of three, smell like almonds, or have milky sap should also be avoided. People should not assume a plant is safe because they see an animal eating it. Many animals have adapted so they can eat harmful plants without suffering.

HOW TO FISH WITH A TRAP

On an island, all it takes are some sticks, patience, and know-how to catch dinner. A fish trap is a fence that allows fishers to trap fish that are swimming in streams. The fence can be made by pushing sticks upright into the mud. It should have a narrow opening, so that fish can swim into the enclosure but then can't find their way out. The sticks should be close together, so that even the smallest fish cannot swim between them. Dead fish and worms can be used to lure fish inside the trap.

BEWARE DANGEROUS ANIMALS

Although many island animals are harmless to humans, others can be deadly. When stranded on an island, it is important to beware of any animals that could cause harm. Danger can be spotted by looking for large footprints or animal droppings, which might be a sign that there are predators nearby.

WATER RISKS

It is best for people to fish only in shallow water, so they can get ashore quickly if they see a shark. It is important to avoid the water once darkness falls. Sharks are most active at night, and it is harder to see them coming. Some animals hunt in murky inland waters, such as rivers and lakes. It is a good idea to avoid walking along riverbanks or through thick vegetation near the water. Crocodiles hide underwater waiting for animals to come nearby. Crocodiles are not very fast on land, so if people can get to shore in time, they may be able to run away.

ANIMAL ATTACK

Seals can be **aggressive**. They may bite, which can lead to infection. There are even stories of seals dragging people out to sea or chasing them. Wild boars will attack if they feel threatened. They will chase people, and once they are near, push them over. Then, they use their tusks to stab the victim in the thighs. Anyone who comes across wild boars should look for a tree to climb, an object to hide behind, or they should back away slowly. Running away will cause the boar to give chase.

BE PREPARED

Animals are attracted to the smell of food. Keeping the campsite clean and free of uneaten food will help make sure that wild animals stay away. The smell of food can cling to clothing. People should also keep any clothes worn while cooking away from the shelter. They can be hung on a tree branch far from the camp, along with any uneaten food.

Wild boars will come to a shelter if they smell food there.

Great white sharks can swim at speeds of up 25 miles per hour (40 km/h).

SURVIVAL STORY

CAPSIZED IN A STORM: JESSE BRILLON

Haida fisherman Jesse Brillon grew up learning about traditional Indigenous peoples' ways of life in British Columbia, Canada. He knew how to live off the land and the sea, just as his ancestors had done. This skill proved especially useful when Jesse's boat capsized in stormy seas on May 9, 2012. Jesse could not avoid the sudden storm to get to safety, but he was able to ride it out. Thanks to his knowledge of the land, he had the survival skills to keep safe until help arrived.

LOST AT SEA

Jesse was fishing for prawns in Hecate Strait off the coast of British Columbia with brothers Brian and Dave Martynuik who are also Haida. When a storm rolled through the area without warning, their boat overturned. The three men were lost at sea. The men had to swim hard to get out from under the boat. It took a few minutes, but everyone was able to climb on top of the capsized boat. They managed to take a tarp and an inflatable raft from the boat. The men had sent out a call to the coast guard for help, but no one had heard it. Their families were not expecting to hear from them for a few weeks, so they would not report the men missing until then. They had to find a way to survive on their own.

RESCUED AT LAST

Jesse, Brian, and Dave huddled together to keep warm as they drifted on the ocean for the next 18 hours. When they spotted an island in the distance, they used the tarp as a sail to guide their raft toward it. They finally washed up on Banks Island. The men put their knowledge of the land to work. They built a shelter from driftwood, and they began gathering food, such as sea urchins, seaweed, and clams. They found a small stream of fresh water to drink. It kept them well hydrated throughout their time on the island. They made a huge mound of garbage to help grab the attention of anyone who might sail or fly by. They all remained hopeful that help would come. Finally, it did. A fishing boat spotted them while sailing along the shore. They were rescued after 10 days on the island.

Driftwood is a valuable material found along the shoreline. It can be used by shipwreck survivors to build shelters and rafts.

BUILD A FIRE

Fire can be used to cook food and boil water. This helps kill any germs and parasites that can make people sick. Fire also provides light and warmth, and can keep away wild animals and insects. It can also be used to signal for help.

KEY INGREDIENTS

There are many ways to start a fire using items found from a wreckage or in nature. The three key ingredients to build a fire are tinder, kindling, and wood. Tinder is the base layer. It can be made from dry grass, dead leaves and pine needles, tree bark, cattail fluff, paper, or anything that will catch fire easily. It needs to be as dry as possible for a spark to burn into a flame. Once the tinder catches fire, it burns out quickly, so kindling is needed to keep it going. Kindling can be made from twigs and branches placed on top of the tinder. Wood keeps the fire burning. Branches and logs can be stacked like a tepee over the kindling.

As well as providing heat and light, fires also provide a welcome comfort to survivors on an island.

START A FLAME

Survivors may not have matches or a lighter to start a fire. Instead, they can use a piece of glass, a mirror, or the bottom of a tin can to direct sunlight onto the tinder. When it gets hot enough, it will start to smoke. Blowing on it gently will cause it to ignite. Another way to start a fire is the fire-plow method. This involves cutting a groove into a piece of softwood with a rock or knife and placing tinder inside the groove. A hard stick is then rubbed quickly back and forth along the groove. The friction, or rubbing force, will create heat, and the tinder will start to smoke.

HOW TO MAKE TINDER

People can make their own tinder by using a knife to shave pieces of wood from a branch. First, they need to remove the bark from the branch. The wood underneath will be drier than the top layer. Then, they can run the knife away from themselves along the branch to make wood curls.

LEARN BASIC FIRST AID

Even minor injuries can get worse over time and threaten the chances of survival. Ships and planes almost always have a first aid kit and some medical supplies on board. They can often be found among the wreckage. If there is no first aid kit, knowledge of basic first aid techniques can mean the difference between life and death.

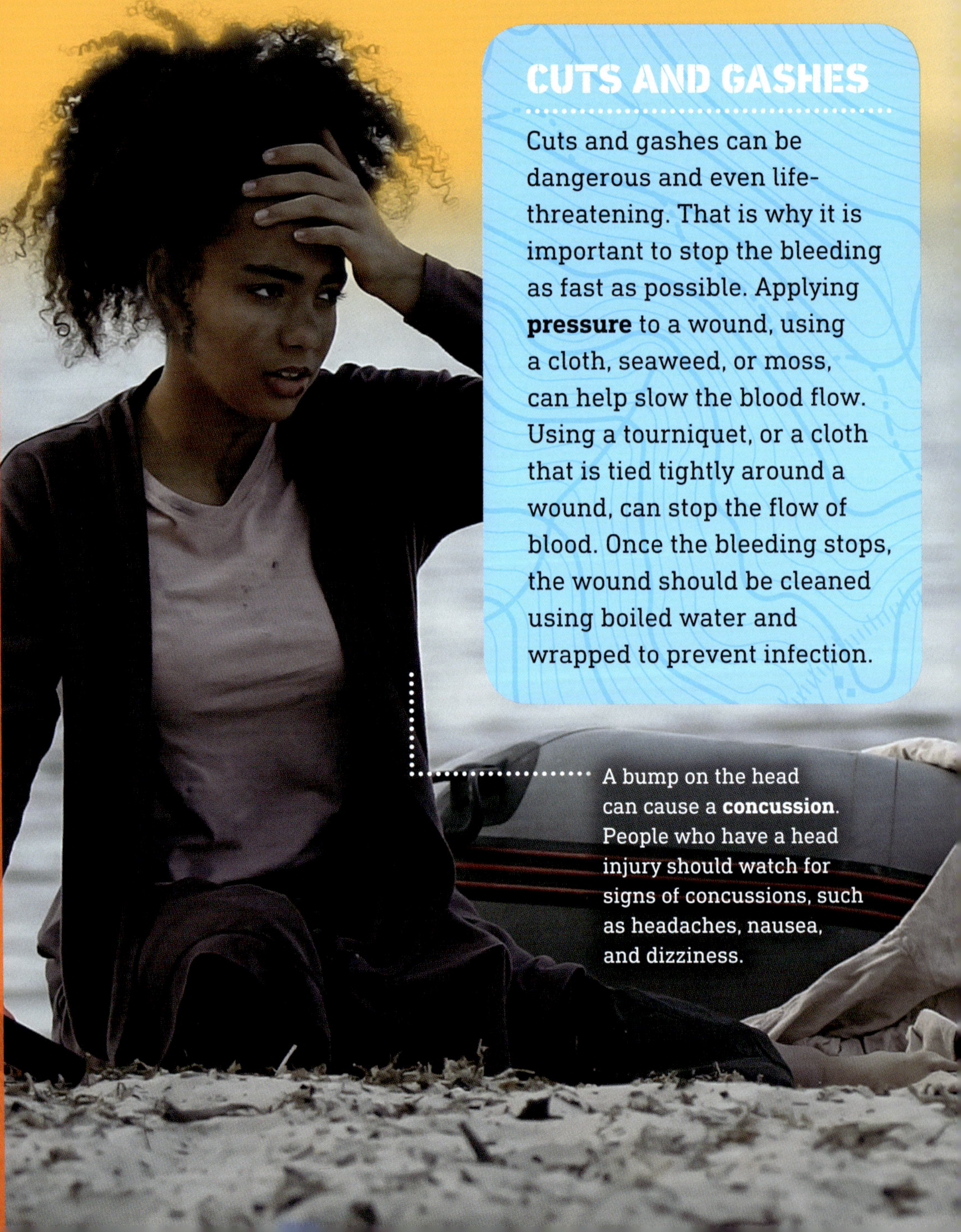

CUTS AND GASHES

Cuts and gashes can be dangerous and even life-threatening. That is why it is important to stop the bleeding as fast as possible. Applying **pressure** to a wound, using a cloth, seaweed, or moss, can help slow the blood flow. Using a tourniquet, or a cloth that is tied tightly around a wound, can stop the flow of blood. Once the bleeding stops, the wound should be cleaned using boiled water and wrapped to prevent infection.

A bump on the head can cause a **concussion**. People who have a head injury should watch for signs of concussions, such as headaches, nausea, and dizziness.

SPRAINS AND BREAKS

It is easy to twist an ankle walking on rocky ground or sprain a wrist by falling. It is important for the injured person to keep the limb still. A **splint** can be made by placing sticks at the joints above and below the injury. Pieces of cloth, rope, vine, or seaweed can be used to tie the sticks in place.

EXTREME WEATHER

In very hot places, **heat exhaustion** is a concern. Staying in the shade and drinking a lot of water is important. In cold places, hypothermia can be an issue. Hypothermia happens when the body loses heat faster than it can produce it. Symptoms include shivering, confusion, and shallow breathing. The key is to warm up by finding shelter that is dry and protected from the wind. Removing wet clothing is also important. Sitting close to the fire and drinking hot water helps, too.

BE PREPARED

Mosquitoes and ticks can carry diseases. Keeping arms and legs covered can help prevent bites. Insect **repellent** can also be used to keep insects away. People can rub mud on their bodies if they do not have repellent on hand. Bugs cannot bite through the mud.

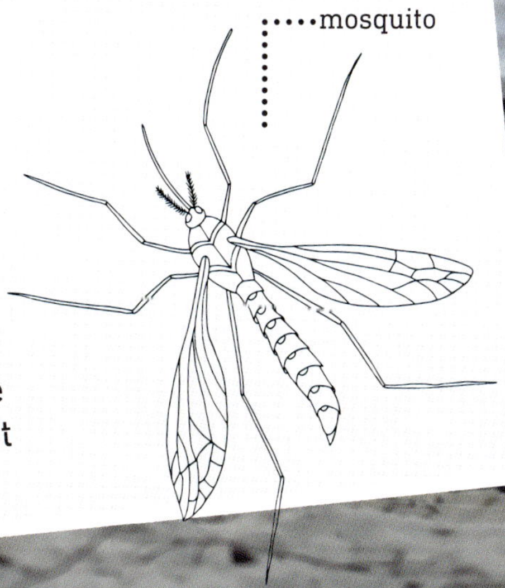

SURVIVAL STORY

HELP!: U.S. NAVY RESCUE

Sometimes even experienced fishermen get swept out to sea. When they do, they need to make smart decisions in order to survive. Three men from a small island in Micronesia were stranded on a deserted island in the Pacific Ocean in April 2016 after a large wave struck their boat. They used their survival skills to help rescuers find them by spelling out HELP on a sandy beach. It was three days before rescuers spotted their sign.

ROUGH WATER

The three fishermen had set out from the island of Pulap in Micronesia, which is northeast of Papua New Guinea. The men hit rough weather, and their boat capsized, forcing them to swim for their lives. They were all wearing life jackets, which helped them stay afloat. They swam for 2 miles (3.2 km) before coming ashore on Fanadik, a deserted island hundreds of miles north of Papua New Guinea.

Life jackets are brightly colored to help make it easier for rescuers to spot people from overhead or far away.

DISTRESS SIGNAL

The men knew they needed to send out a distress signal to let people know where they were. They collected palm leaves to write the word "HELP" in big letters in the sand. Then, they built a fire and waited on the beach. They knew their families would be worried and report them missing when they did not arrive at their destination. It was only a matter of time before someone found them.

HELP ARRIVES

The U.S. Coast Guard got word of the missing men on April 5 and immediately began to search for them. The Coast Guard alerted the local military and other boats in the area to be on the lookout for the fishermen. Planes from a nearby U.S. Navy base in Guam joined the search.

Finally, on April 7, a Navy pilot saw a strange image on his **infrared camera**. It had picked up heat from the fire that the men had built, even though it had burned out. The pilot flew around the island and saw the men's sign on the beach. Within seconds, the pilot spotted the men running out of the forest. They were waving their life jackets and shouting. The plane dropped a smoke signal to let the men know they had been found. Soon after, a local boat arrived to rescue the men. It took them back home to Pulap, which was only about 3 miles (5 km) from the island where they had been stranded.

Staying alive until help arrives is the only thing that matters when stranded on an island. Survivors should take a good look at their surroundings and come up with a plan before taking any action. Then, they can put their survival skills to use to find water, build a shelter, look for food, and make a fire.

BE CAREFUL

Any activities that could put people in danger or cause injury should be avoided. This includes keeping away from deep water where big fish are more likely to swim and staying away from caves where dangerous animals, such as eels, might live. Getting plenty of rest is also important. This will help survivors stay alert for hazards and watch for any sign of help. Staying in one place and waiting to be rescued is the best option. Rescuers are more likely to look for people near the wreckage. Ships and planes send back information about their location that can help rescuers find them if they are in trouble. There is also more risk of injury when exploring a new area.

moray eel

HOW TO MAKE A RESCUE SIGNAL

The best way for survivors to be spotted on an island is to make a rescue signal. They can use branches, leaves, stones, and any bright or reflective objects from the wreckage to spell **SOS** or HELP on the beach. The letters need to be big enough for aircraft to spot them easily as they fly overhead. Building a large fire on the beach will help draw the attention of passing planes. Adding green leaves to the flames will create more smoke that can be seen from farther away.

STAY POSITIVE

It is important to remain calm and positive. Panicking wastes valuable energy. It also leads to poor decision-making that could put a person's life at greater risk. Complaining or breaking down will not change the situation. Thinking happy thoughts is a good way for survivors to stay positive at all times.

GLOSSARY

GLOSSARY

adapt Develop skills or physical features over time to help animals live in a certain biome

aggressive Describes behavior that is angry or violent

bacteria Single-celled organisms that are invisible to the human eye

capsize Turn upside down in the water

charter Book for private travel

concussion A brain injury resulting from a blow to the head

dehydration Excessive loss of water from a person's body

desolate Describes a place that feels empty and where no one lives

exposure Being affected by something, such as the Sun

fiber A thread or strand from a plant or other natural thing

flares Lights that shine brightly for a short time; used in emergencies

fresh water Water found in streams, ponds, rivers, and waterfalls that is not salty

Indigenous peoples The original inhabitants of a land

hazards Dangrous objects

heat exhaustion A medical condition that results from the body getting too hot, in which a person may feel dizzy, thirsty, and weak

hypothermia When a person's body temperature becomes too low, causing serious damage

infrared camera A camera that detects heat and converts it into an electric signal to create an image

inhabited Lived in or occupied

lean-to A simple structure made by leaning a branch against a tree and covering it with other branches and leaves

navigation The process of finding one's way

parasites Organisms that live and feed on other animals

predators Animals that hunt and eat other animals

pressure A constant force

privateer A sailor hired by a government to raid enemy ships

remote Describes something that is far away from other things

repellent A substance that keeps insects away

salvage Save

SOS A universal rescue signal that stands for Save Our Souls

splint A hard object, such as a piece of wood, that holds a broken bone in place

tepee A portable cone-shaped tent made of cloth, canvas, or skins

tripod A stand that has three legs

tropical Having a hot, humid climate; places that are near the equator

LEARNING MORE

LEARNING MORE

Find out more about how to survive on an island.

Bailey, Diane. *Survival Skills* (The Great Outdoors). Mason Crest, 2016.

Bailey, Gerry. *Stranded on an Island* (Science to the Rescue). Crabtree Publishing, 2014.

Gifford, Clive. *Ultimate Guide to Surviving in the Wild*. Frances Lincoln Children's Books, 2019.

Kopp, Megan. *Islands Inside Out* (Ecosystems Inside Out). Crabtree Publishing Company, 2015.

WEBSITES

Learn how to start a fire without a match:
https://adventure.howstuffworks.com/survival/wilderness/how-to-start-a-fire.htm

Find out more about wilderness survival at:
www.backpacker.com/survival

Discover some of the world's islands at:
www.touropia.com/uninhabited-islands-around-the-world

Get to know basic first aid procedures at:
https://troopresources.scouting.org/first-aid-skill-activities/

INDEX

ABOUT THE AUTHOR

Heather C. Hudak has written hundreds of books for children. When she's not writing, Heather loves to travel the world. She has been to all kinds of biomes, from deserts to mountains and everything in between. Heather has visited many beautiful remote islands in Asia and Central America, but she hopes she never gets stranded on one!